# BULL SHARK

**By Rachel Rose**

Consultant: Erin McCombs
Educator, Aquarium of the Pacific

Minneapolis, Minnesota

**Credits**

Cover and title page, © Fiona Ayerst/Shutterstock, © Ton Anurak/Shutterstock, © AP focus/Shutterstock; 3, © nabil refaat/Shutterstock; 4-5, © Michael Patrick O'Neill/Alamy; 6-7, © pefcon/iStockPhoto; 8-9, © Stocktrek Images, Inc./Alamy; 10-11, © Dmitry Vinogradov/Getty; 12-13, © Gérard Soury/biosphoto; 14-15, © Stocktrek Images, Inc. /Alamy; 16-17, © Media Drum World/Alamy; 18-19, © Cultura Creative RF/Alamy; 20-21, © Martin Prochazkacz/Shutterstock; 22, © Carlos Grillo/Shutterstock; 22-23, © nabil refaat/Shutterstock; 24, © nabil refaat/Shutterstock.

**President:** Jen Jenson
**Director of Product Development:** Spencer Brinker
**Senior Editor:** Allison Juda
**Associate Editor:** Charly Haley
**Designer:** Colin O'Dea

*Library of Congress Cataloging-in-Publication Data*

Names: Rose, Rachel, 1968- author.
Title: Bull shark / by Rachel Rose.
Description: Minneapolis, Minnesota : Bearport Publishing Company, [2022] | Series: Shark shock! | Includes bibliographical references and index.
Identifiers: LCCN 2021026706 (print) | LCCN 2021026707 (ebook) | ISBN 9781636915296 (library binding) | ISBN 9781636915388 (paperback) | ISBN 9781636915470 (ebook)
Subjects: LCSH: Bull shark--Juvenile literature.
Classification: LCC QL638.95.C3 R654 2022 (print) | LCC QL638.95.C3 (ebook) | DDC 597.3/4--dc23
LC record available at https://lccn.loc.gov/2021026706
LC ebook record available at https://lccn.loc.gov/2021026707



For more information, write to Bearport Publishing, 5357 Penn Avenue South, Minneapolis, MN 55419. Printed in the United States of America.

# CONTENTS

# Bump and Bite

A bull shark hunts in **murky** ocean waters along the coast. With its strong sense of smell, it sniffs out a large fish. The bull shark bumps its head against the fish. Then, *wham!* It clamps down with a massive bite. Dinner is served!

Bull sharks bump into their **prey** to find out what they're about to bite.

# Fresh and Far Out

Like most sharks, bull sharks live in salty ocean waters around the world. But bull sharks have **adapted** to other waters, too. They are the only ocean sharks that also **survive** in freshwater rivers and lakes!

And these sharks do more than just dip their tails into freshwater. They have been found as far as 2,500 miles (4,000 km) up the Amazon River in South America.

Bull sharks hold salt inside their bodies when they are in fresh water.

# BULL SHARKS AROUND THE WORLD

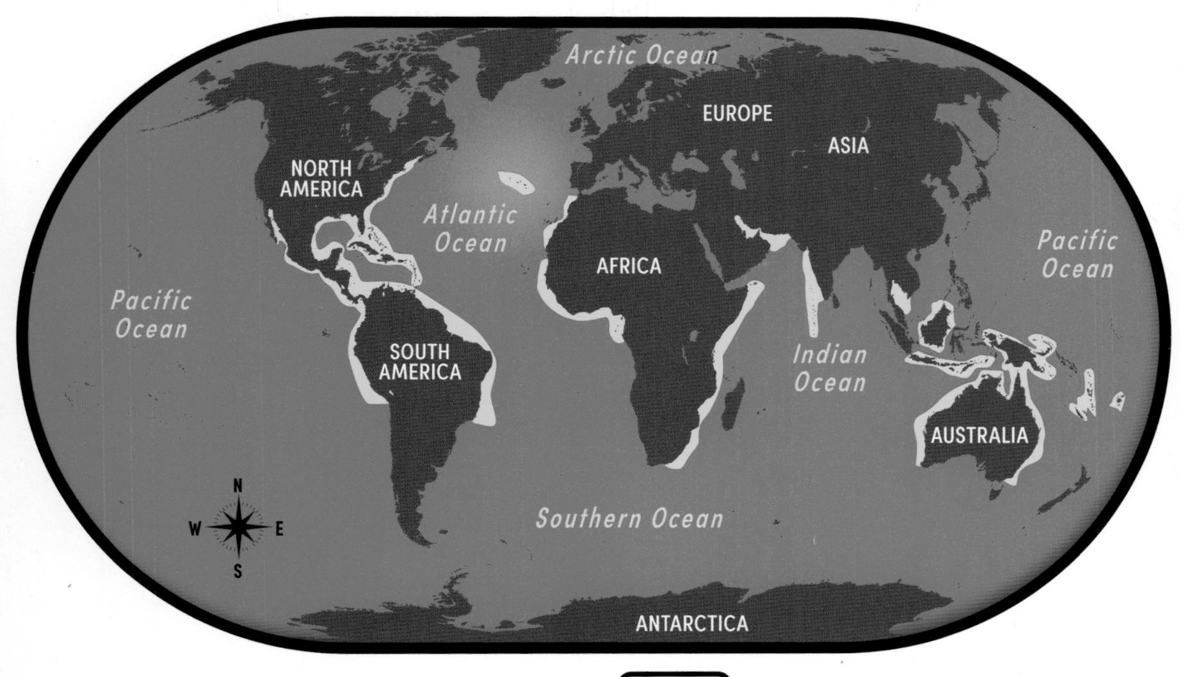


Where **bull sharks live**

# Bull Headed

Similar to the bulls they are named after, these sharks have big heads and wide, flat noses. Also like bulls, they have **bulky** bodies. Bull sharks grow to be about 11 feet (3 m) long. That's longer than the height of a basketball hoop! **Female** bull sharks are usually bigger than **males**.

Although bull sharks have big heads, their eyes are small. They can't see well.

# SNEAKY HUNTERS

These big sharks often hunt in murky waters where other fish can't see them coming. Their coloring helps them blend in so they can sneak up on their prey. When seen from below, bull sharks' white bellies look like sunlit waters near the surface. From above, their gray backs blend in with deeper, darker waters.

Bull sharks have an excellent sense of smell, which helps them find their prey in murky waters.

# Strong Bite!

A bull shark is more than just sneaky—it also has the strongest bite of any shark! The **force** of its bite is more than four times stronger than a dog's bite. And the bull shark's bite is full of several rows of very sharp teeth. When teeth in the front rows fall out, they are replaced by the teeth behind them.

A bull shark has about 350 teeth in its mega mouth!

# What's on the Menu?

A superstrong bite means that bull sharks can eat just about anything. When they are in the ocean, they usually eat smaller fish. Sometimes, they hunt dolphins, turtles, and other sharks. In rivers and lakes, bull sharks might eat shrimp and crabs. They have even been known to chomp into hippos!

Bull sharks hunt for food at any time of day or night.

# The Hunter Is Hunted

Even though bull sharks are strong hunters, they still have a few **predators**. Sometimes, these sharks are hunted by larger sharks. When they swim in rivers, they can also be attacked by crocodiles!

However, the biggest **threat** to bull sharks is humans. People hunt and kill bull sharks for their skin, fins, and meat.

Bull sharks spend a lot of time near coasts. So, they often get caught and injured by fishing nets.

A crocodile eating a bull shark

# Pup Time

Bull sharks spend most of their time alone. These sharks only come together to **mate**. This often happens in freshwater areas during the late summer.

About 11 months after mating, it is time for females to give birth. Bull shark pups are usually born in lakes or river mouths, where there are fewer predators. This keeps the pups safe.

Bull shark mothers can have up to 13 pups at a time.

# Out to Sea

As soon as the pups are born, their mothers leave them. Most pups stay in rivers or lakes until they are fully grown. Then, they are ready to head out into the ocean to find mates and have pups of their own.

Bull sharks usually live for about 30 years.

# MORE ABOUT BULL SHARKS

In 1937, a bull shark was found in the Mississippi River in Illinois. It swam thousands of miles from the Gulf of Mexico!

The largest bull shark ever found was about 13 ft (4 m) long.

The bull shark is also known by many other names, including river shark and square-nose shark.

Bull sharks often live in water that's less than 100 ft (30.5 m) deep.

Bull sharks weigh between 200 and 500 pounds (90 and 230 kg). That's about as heavy as a lion!

# Glossary

**adapted** changed over time to survive in certain surroundings

**bulky** thick and wide

**female** a bull shark that can give birth to young

**force** power or strength

**males** bull sharks that cannot give birth to young

**mate** to come together to have young

**murky** cloudy or not very clear

**predators** animals that hunt other animals for food

**prey** animals that are hunted by other animals

**survive** to stay alive

**threat** something that could cause harm

# Index



# Read More

**Alderman, Christine Thomas.** *Bull Sharks (Bolt: Swimming with Sharks).* Mankato, MN: Black Rabbit Books, 2020.

**Nixon, Madeline.** *Bull Shark (Sharks).* New York: AV2, 2019.

# Learn More Online

1. Go to **www.factsurfer.com** or scan the QR code below.
2. Enter "**Bull Shark**" into the search box.
3. Click on the cover of this book to see a list of websites.

# About the Author

Rachel Rose lives in California. She swims in the ocean every day, and she sees plenty of seals—but she hasn't seen a shark yet!